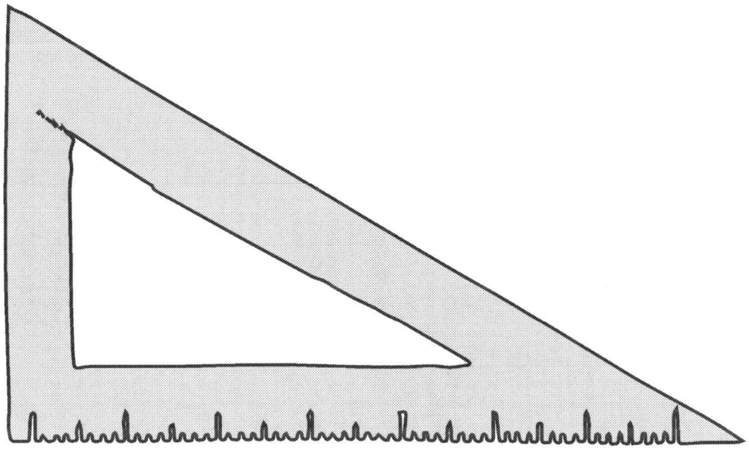

This Notebook Belongs To:

Subject:

Grade:

Made in the USA
Las Vegas, NV
22 January 2021

16242500R00061